Magical Elemental Atoms

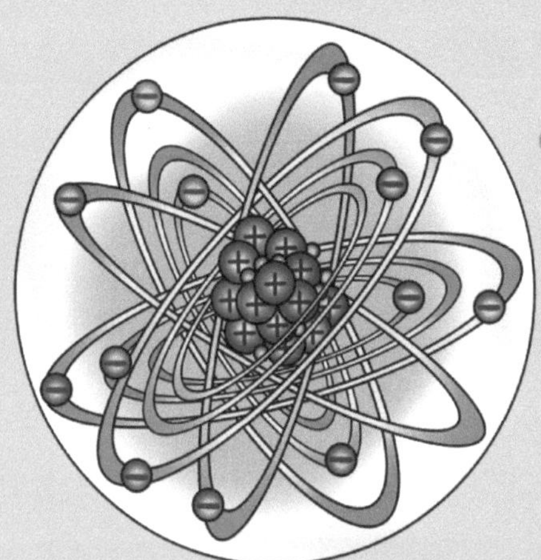

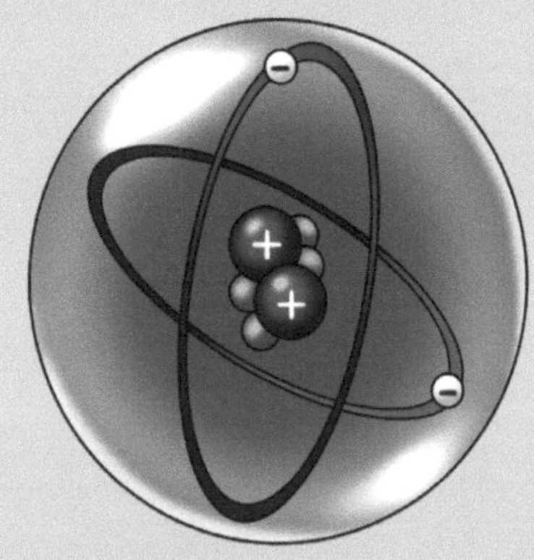

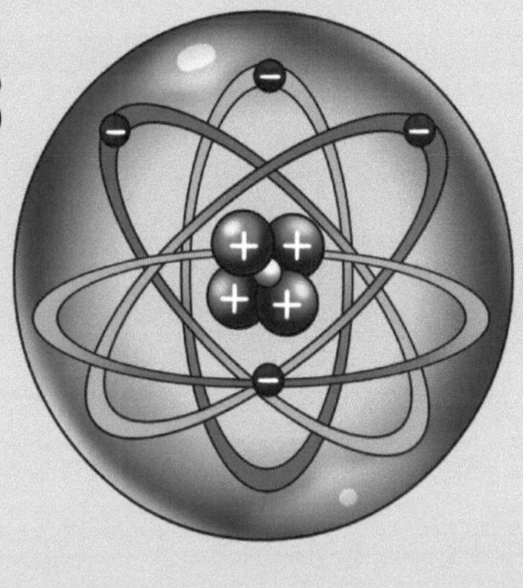

Count the Protons and Electrons

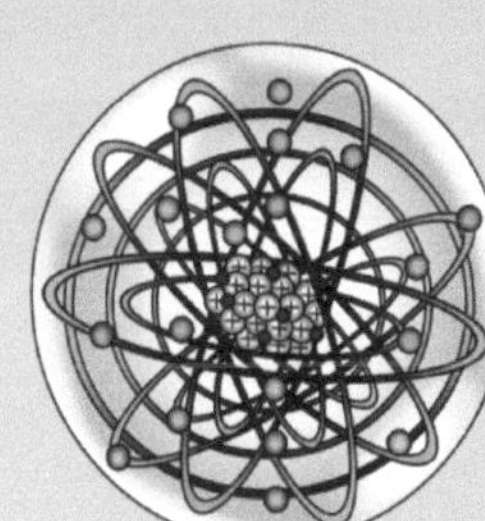

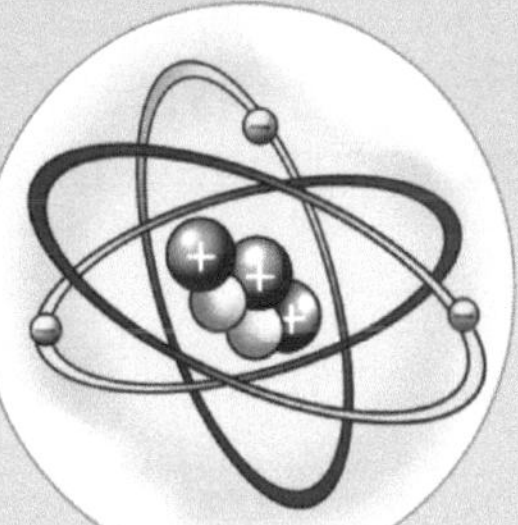

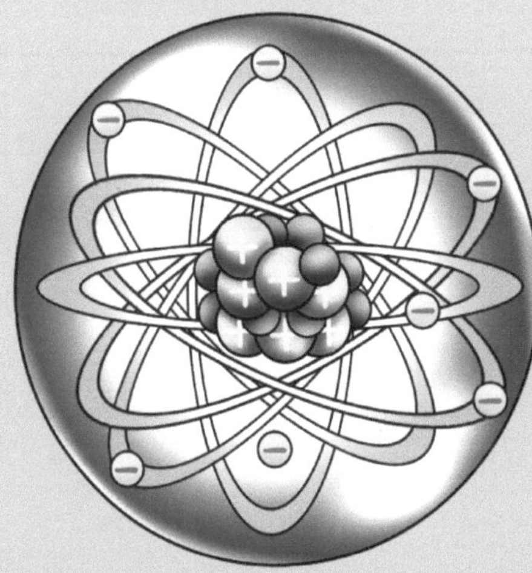

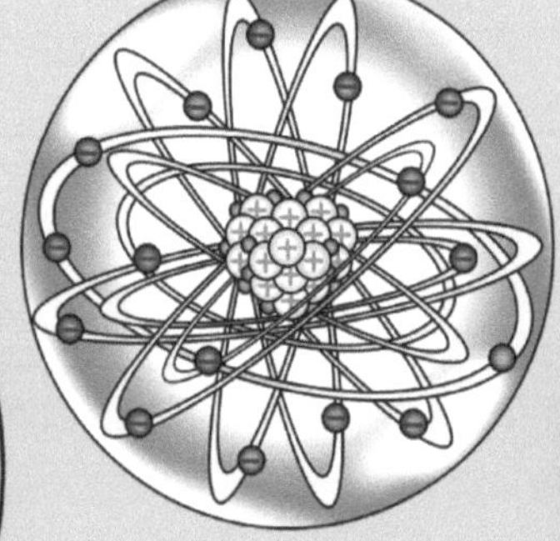

By Sybrina Durant with Illustrations By Pranava

Magical Elemental Atoms

Count the Protons and Electrons

Soft Cover: ISBN: 978-1-942740-62-9

Copyright 2026

BISAC Codes:

JNF051050 JUVENILE NONFICTION / Science & Nature / Earth Sciences / General

JNF016000 JUVENILE NONFICTION / Curiosities & Wonders

JNF001010 JUVENILE NONFICTION / Activity Books / Coloring

Soft Cover: ISBN: 978-1-942740-62-9

Hard Cover: ISBN: 978-1-942740-63-6

Magical Elemental Atoms

All periodic table elements are made up of atoms.

All elements, in their neutral atomic state,
have an equal number of protons and electrons.

In a normal state, atoms have the same number of positive protons and negative electrons, which makes their overall charge zero. Protons, which are +1 charged, tell us what kind of element it is, while electrons, which are -1 charged, balance this charge. This one-to-one relationship is very important for the atom to be neutral.

An atom is called "neutral" when it has no overall charge. To be balanced, there need to be the same number of protons and electrons. For instance, if an atom has 10 protons, it also needs 10 electrons to remain neutral.

The atomic number, shown on the periodic table, tells us how many protons are in an atom. For a neutral atom, this number also shows how many electrons it has. For example, a neutral iron atom has 26 protons and 26 electrons.

As stated, all elements have the same number of protons and electrons. But an element can also become an ion, which means the number of protons and electrons is not the same. This happens when an element gains or loses electrons during chemical reactions. For example, some metals, like those in the first two groups of the periodic table, often lose electrons, which makes them positively charged. These are called

Magical Elemental Atoms

(Continued)

cations. On the other hand, some non-metals can gain electrons, making them negatively charged, which are called anions. Basically, any element can have a different number of electrons and protons when it is not electrically neutral.

While protons and electrons must be equal for an atom to be neutral, the number of neutrons can change. Isotopes are atoms that belong to the same element. They have the same number of protons and electrons, but their neutron counts are different. This means some isotopes can be heavier or lighter because of this difference. For example, carbon-12 and carbon-14 are both types of carbon. They each have six protons and six electrons, but carbon-12 has six neutrons and carbon-14 has eight.

The number of protons and electrons in an atom defines what element it is and affects how it reacts with other atoms. So, the chemical properties of isotopes in the same element are very similar. However, because isotopes have different neutron counts, they can have different weights and physical properties, such as density and stability.

Some isotopes are stable, meaning they don't break down over time, while others are unstable and will change into different atoms through a process called radioactive decay. This decay is important for things like medicine and dating ancient objects, like using carbon-14 to figure out how old something is.

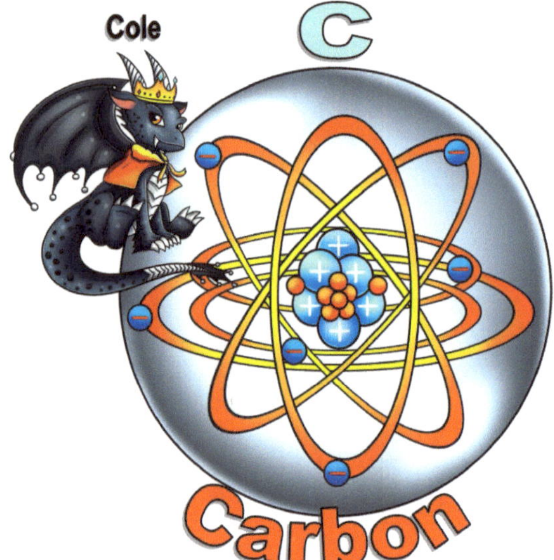

You just learned a lot about atoms but in this book, we are concentrating on the "neutral" atomic state. Count the electrons and protons on the following pages. It will help you remember how many of each make up each periodic table element.

(Begin The Count)

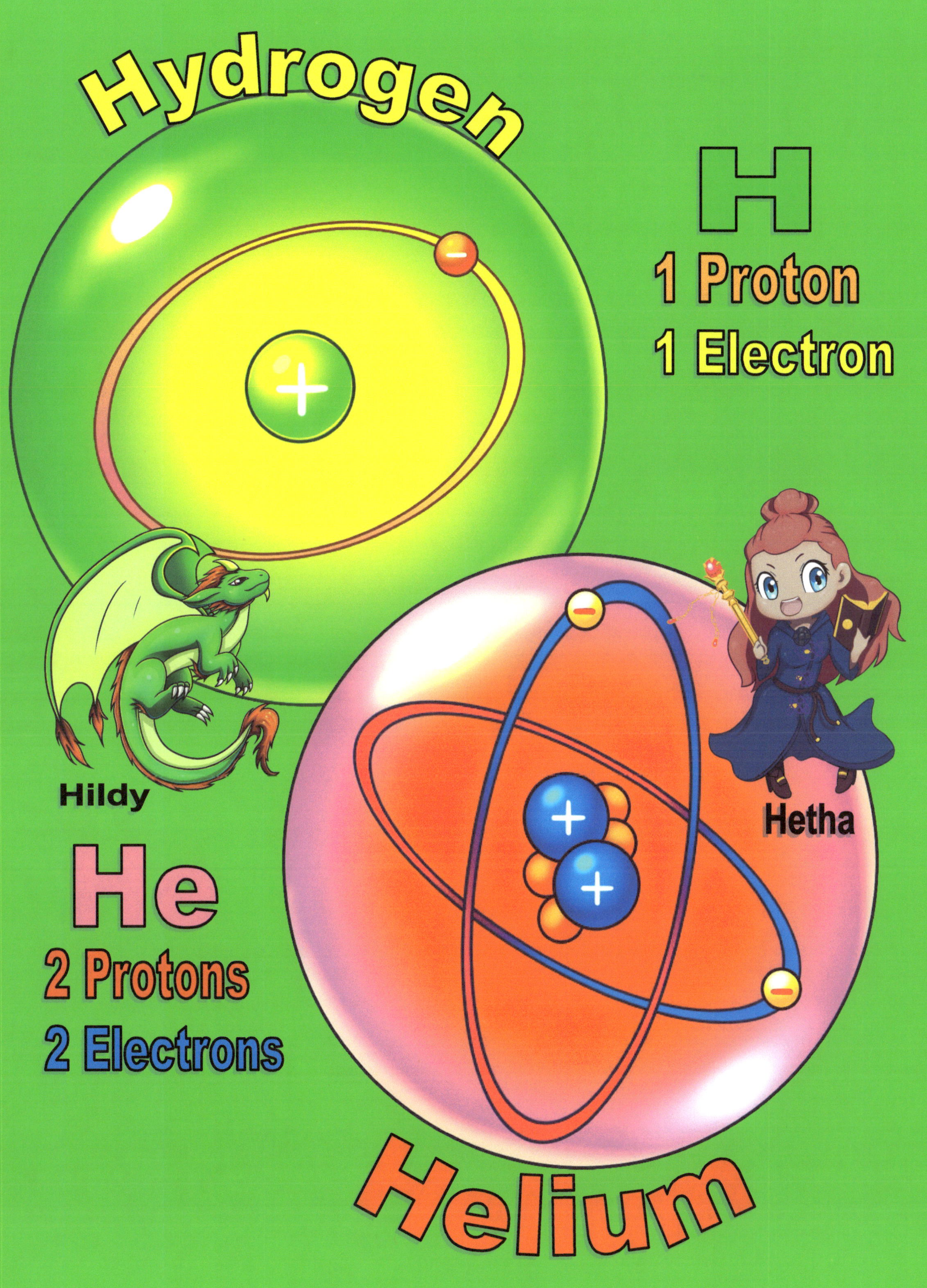
Hydrogen
H
1 Proton
1 Electron
Hildy
He
2 Protons
2 Electrons
Hetha
Helium

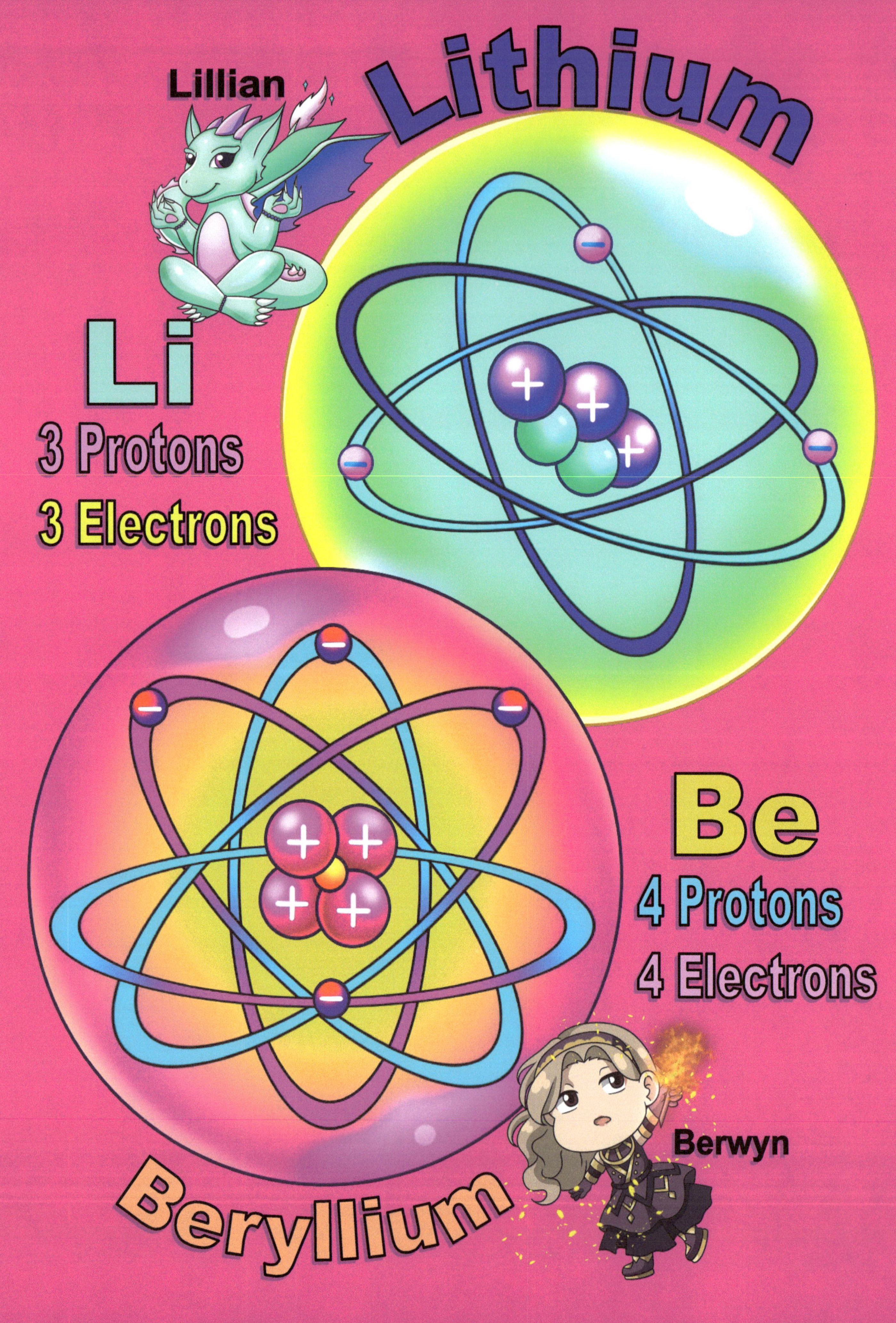

Lillian
Lithium
Li
3 Protons
3 Electrons
Be
4 Protons
4 Electrons
Beryllium
Berwyn

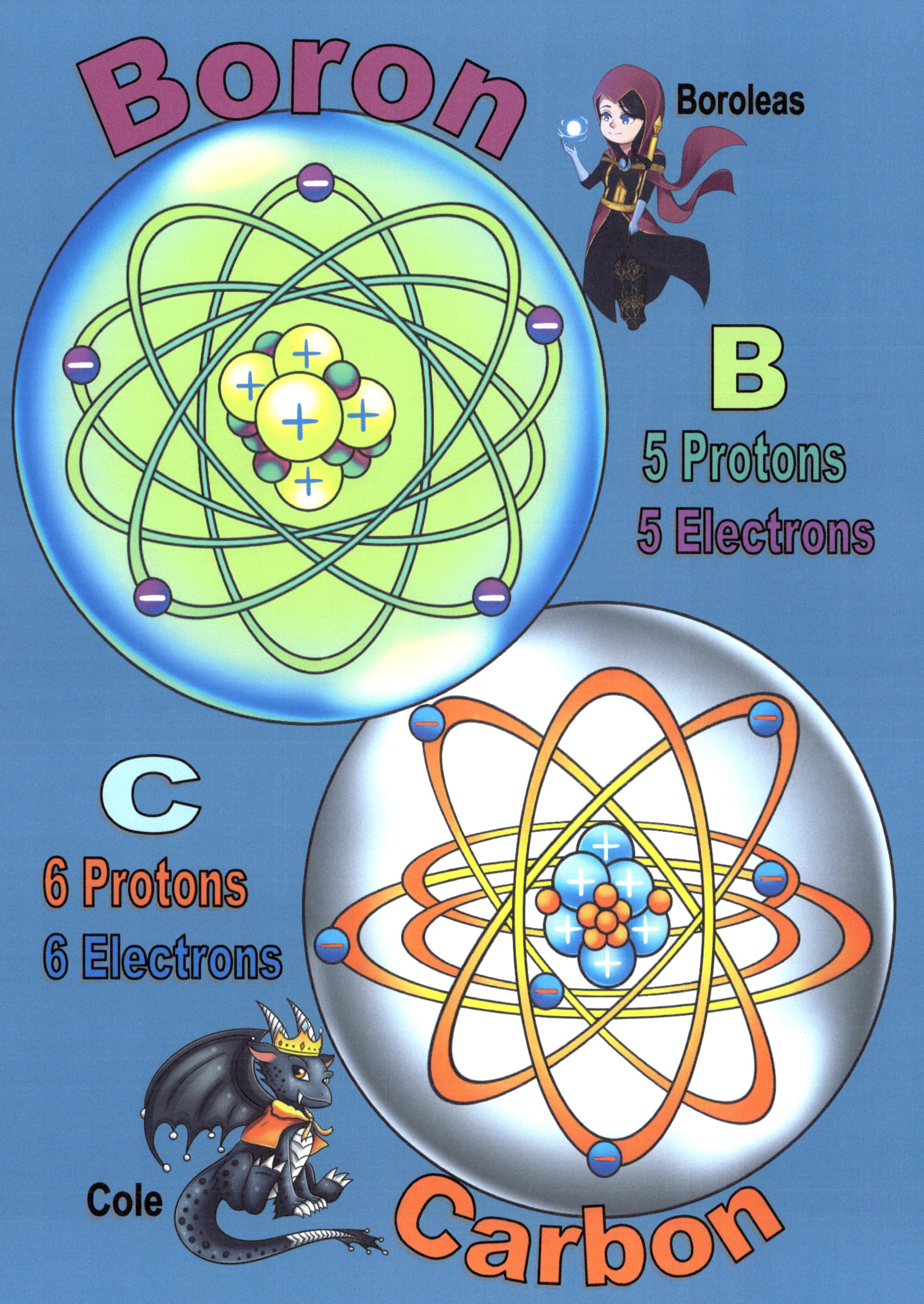

Boron
Boroleas
B
5 Protons
5 Electrons
C
6 Protons
6 Electrons
Cole
Carbon

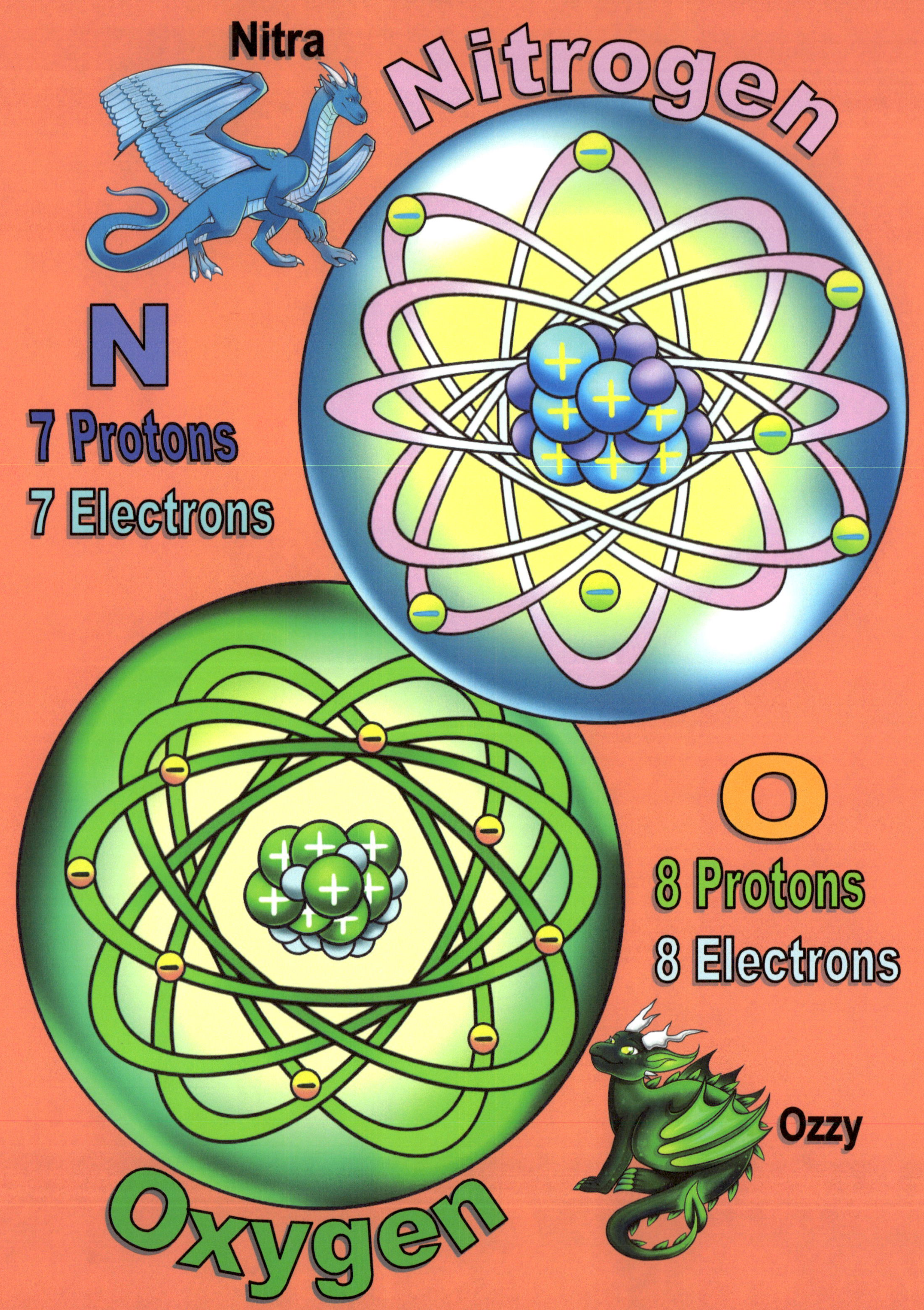
Nitra
Nitrogen
N
7 Protons
7 Electrons
O
8 Protons
8 Electrons
Ozzy
Oxygen

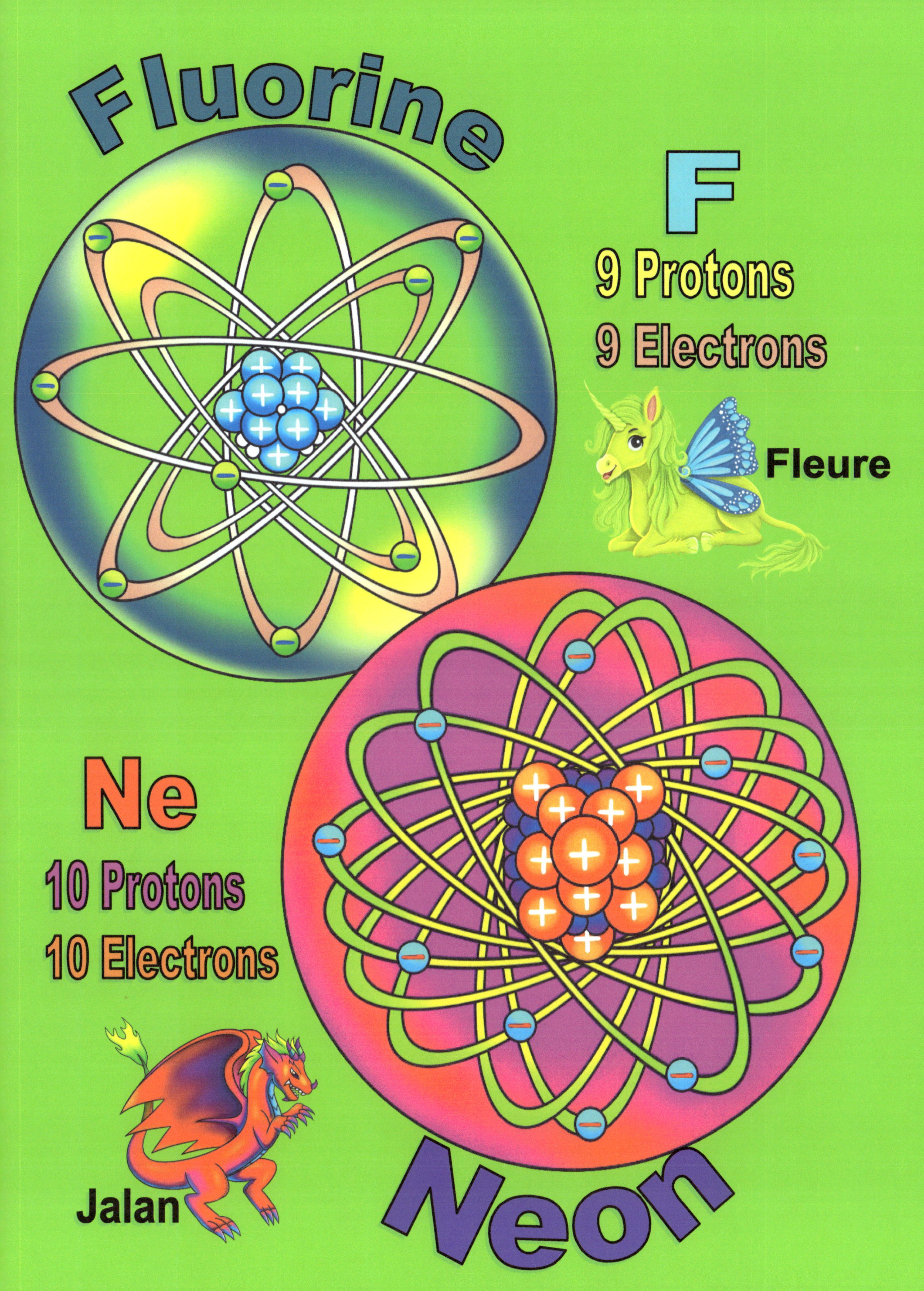
Fluorine
F
9 Protons
9 Electrons
Fleure
Ne
10 Protons
10 Electrons
Jalan
Neon

Sodium
Na
11 Protons
11 Electrons
Sorn

Magnesium
Mg
12 Protons
12 Electrons
Maggie

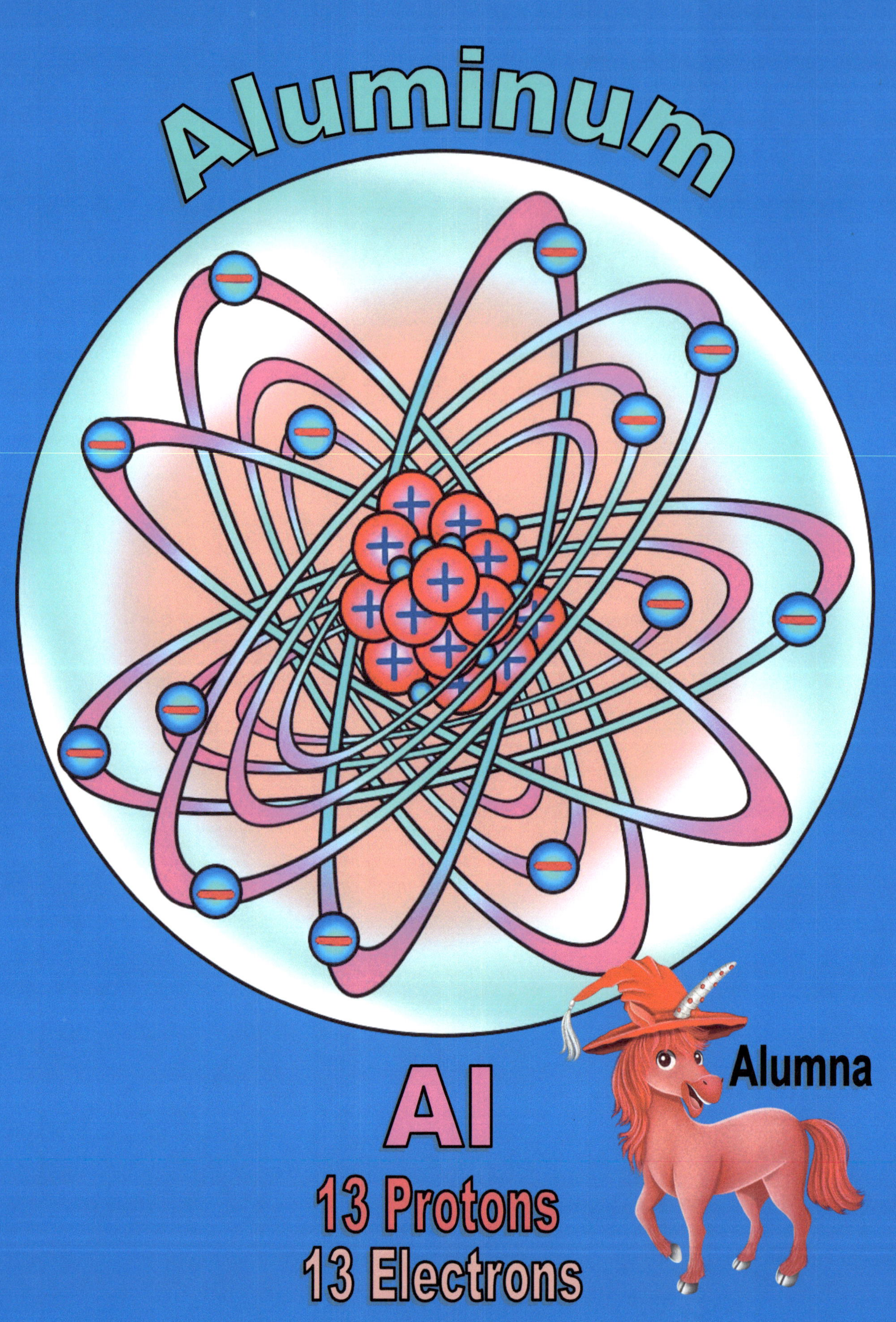

Aluminum
Al
13 Protons
13 Electrons
Alumna

Silicon
Si
14 Protons
14 Electrons
Silonar

Phosphorus
P
15 Protons
15 Electrons
Phova

Sulfur
S
16 Protons
16 Electrons
Xoe

Chlorine
Cl
17 Protons
17 Electrons
Krystix

Argon
Ar
18 Protons
18 Electrons
Areg

Potassium
K
19 Protons
19 Electrons
Pearl

Calcium
Ca
20 Protons
20 Electrons
Verly

Elements 21 thru 118

Instead of showing all the protons and electrons in all atoms, the next pages will have small spheres with 1 or 2 letter abbreviation labels for each element along with a number called the atomic number. The atomic number tells us how many protons and electrons are in that element. This way of showing the rest of the elements is helpful because, when the atomic number is over 20, there are just too many protons and electrons to easily show or count.

You might notice that most element symbols (letter abbreviations) are short forms that make sense, but some just don't. For example, the symbol for Potassium is K, because it comes from an old Latin word, "kalium." This name is linked to ash from plants because potassium was first found in potash. The name kalium helps avoid mix-ups with other elements.

Here are some other unusual element symbols:

Sb for Antimony: Sb is for "stibium," the Latin name for the mineral Antimony comes from. This mineral was often used in makeup and medicine in the early days.

Au for Gold: The symbol Au is from the Latin word "aurum," which means gold. In ancient Rome, gold was called aurum.

Pb for Lead: The symbol Pb is from "plumbum," the Latin word for lead, which relates to plumbing.

Ag for Silver: Ag comes from "argentum," the Latin word for silver. This word has roots in Sanskrit and means "shining."

Na for Sodium: The symbol Na is from "natrium," the Latin name for sodium. In English, we say "sodium," but the symbol reflects its historical name.

Sn for Tin: Sn comes from "stannum," the Latin word for tin and a name for an ancient mixture of silver and lead.

W for Tungsten: The letter W is used for tungsten because it was originally called wolfram. This name is from a mineral that was noticed to consume tin during heating, leading German smelters to call it "wolf's foam."

It could be that it might be easier to remember these particular atomic symbols simply because they are so different from the rest of them.

Enjoy!

Sc
21
Scandra
Scandium
Ti
22
Tilly
Titanium
V
23
Vana
Vanadium
Cr
24
Crowmist
Chromium
Mn
25
Mangar
Manganese
Fe
26
Iown
Iron

Coriss
Co
27
Cobalt
Nix
Ni
28
Nickel
Cu
29
Cuprum
Copper
Zn
30
Dr. Zinko
Zinc
Ga
31
Gallant
Gallium
Ge
32
Gemel
Germanium

As
33
Arkyn
Arsenic
Se
34
Selenice
Selenium
Br
35
Brogach
Bromine
Kr
36
Krypto
Krypton
Rb
37
Ruby
Rubidium
Sr
38
Strauna
Strontium

Yago
Y
39
Yttrium
Zora
Zr
40
Zirconium
Nonnach
Nb
41
Niobium
Maximo
Mo
42
Molybdenum
Tephen
Tc
43
Technetium
Ruth
Ru
44
Ruthenium

Rovana
Rh
45
Rhodium
Paedin
Pd
46
Palladium
Silubhra
Ag
47
Silver
Cadmus
Cd
48
Cadmium
Iker
In
49
Indium
Tinam
Sn
50
Tin

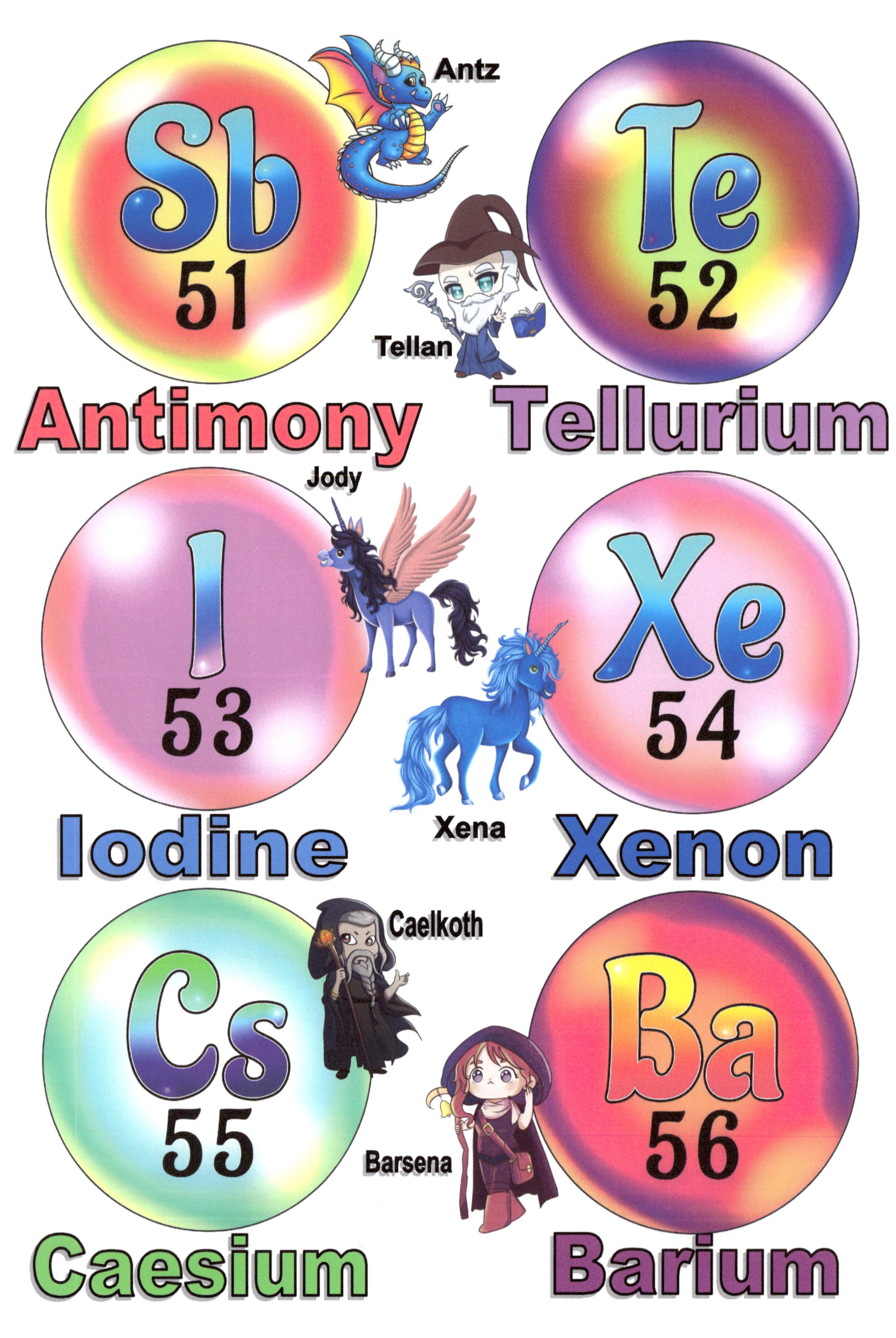
Antz
Sb
51
Antimony
Tellan
Te
52
Tellurium
Jody
I
53
Iodine
Xena
Xe
54
Xenon
Caelkoth
Cs
55
Caesium
Barsena
Ba
56
Barium

Lannion
La
57
Lanthanum
Cerelia
Ce
58
Cerium
Praethec
Pr
59
Praseodyium
Nd
60
Nelushia
Neodymium
Proctor
Pm
61
Samarinda
Promethium
Sm
62
Samarium

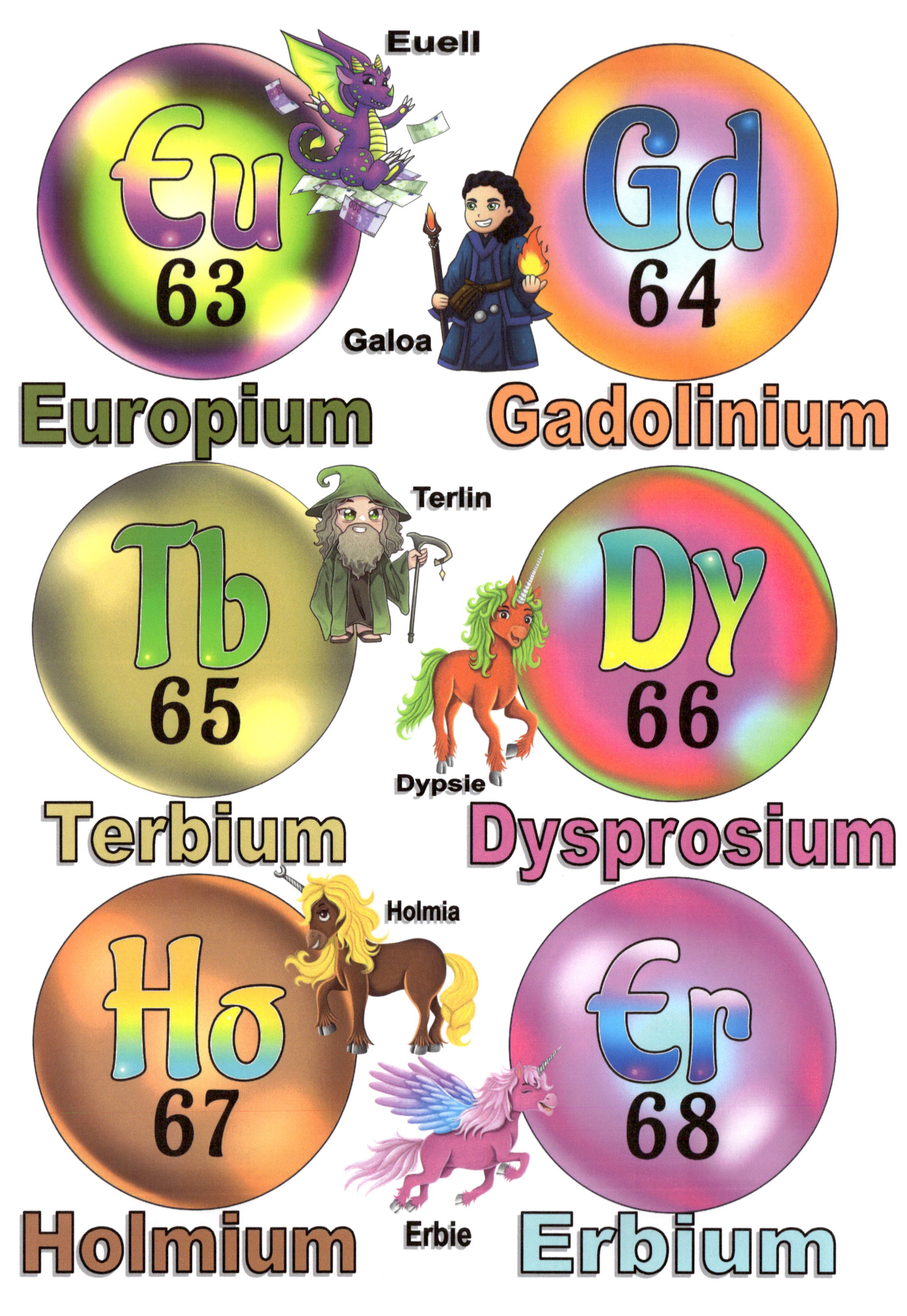

Euell
Eu
63
Europium
Galoa
Gd
64
Gadolinium
Terlin
Tb
65
Terbium
Dypsie
Dy
66
Dysprosium
Holmia
Ho
67
Holmium
Erbie
Er
68
Erbium

Thurwin
Tm
69
Thulium
Yitzy
Yb
70
Ytterbium
Urmi
Lu
71
Lutetium
Hallam
Hf
72
Hafnium
Taltra
Ta
73
Tantalum
Wolfie
W
74
Tungsten

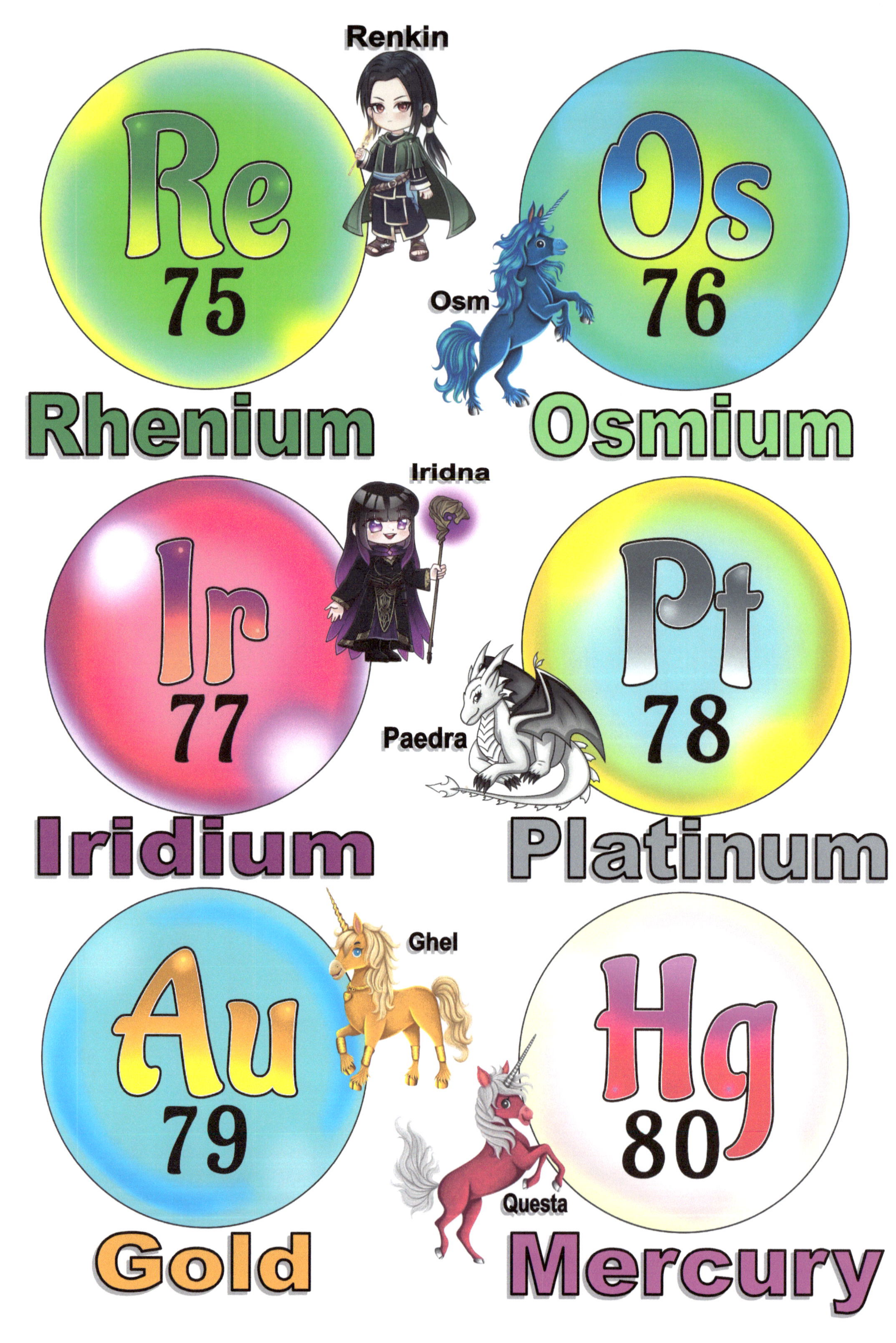

Renkin
Re
75
Rhenium
Osm
Os
76
Osmium
Iridna
Ir
77
Iridium
Paedra
Pt
78
Platinum
Ghel
Au
79
Gold
Questa
Hg
80
Mercury

Thanelen
Tl
81
Thallium
Lauda
Pb
82
Lead
Bitsy
Bi
83
Bismuth
Polgarn
Po
84
Polonium
Aszrad
At
85
Astatine
Ramoran
Rn
86
Radon

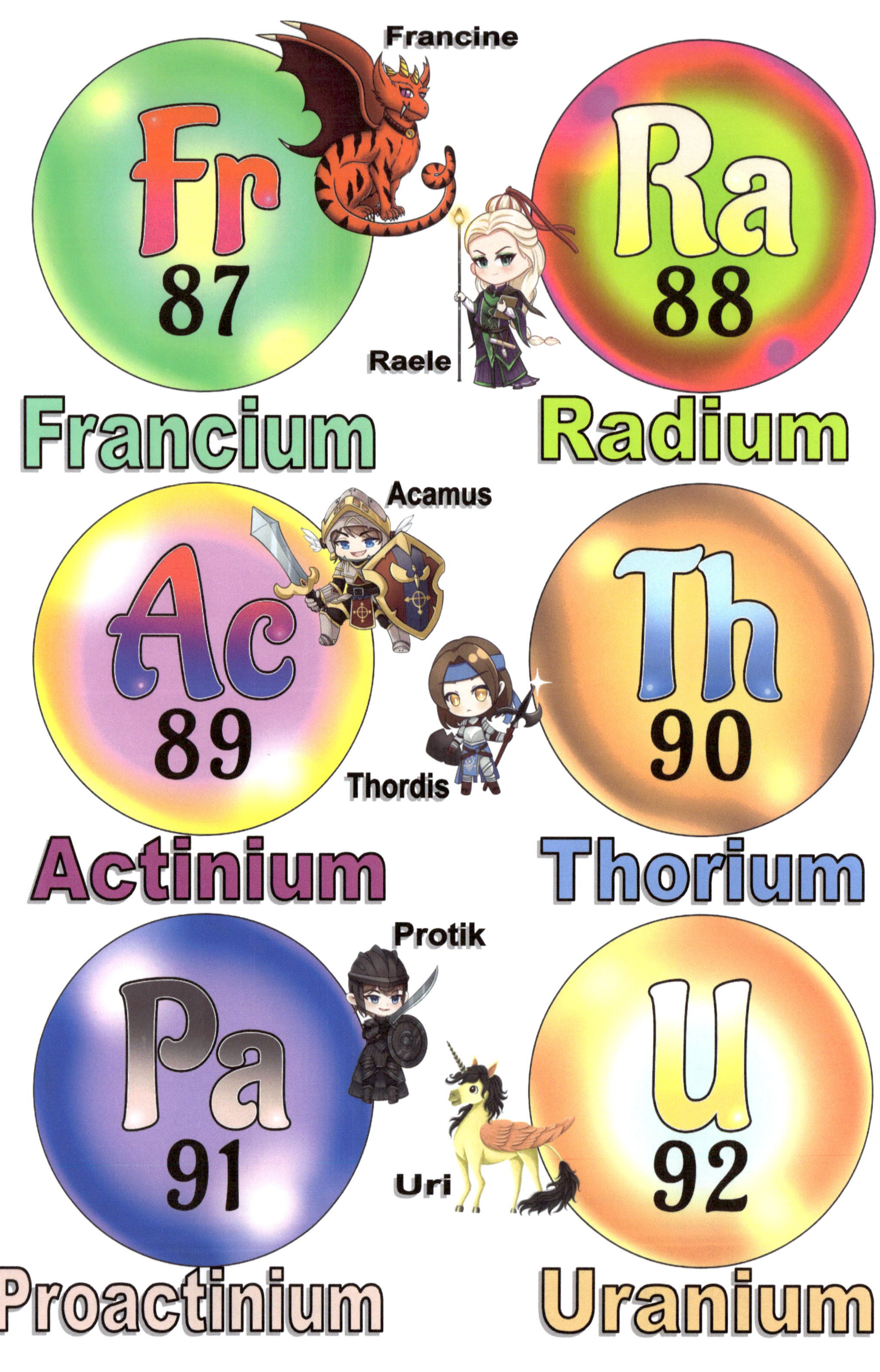

Francine
Fr
87
Francium
Raele
Ra
88
Radium
Acamus
Ac
89
Actinium
Thordis
Th
90
Thorium
Protik
Pa
91
Proactinium
Uri
U
92
Uranium

Neptunium

Plutonium

Americium

Curium

Berkelium

Californium

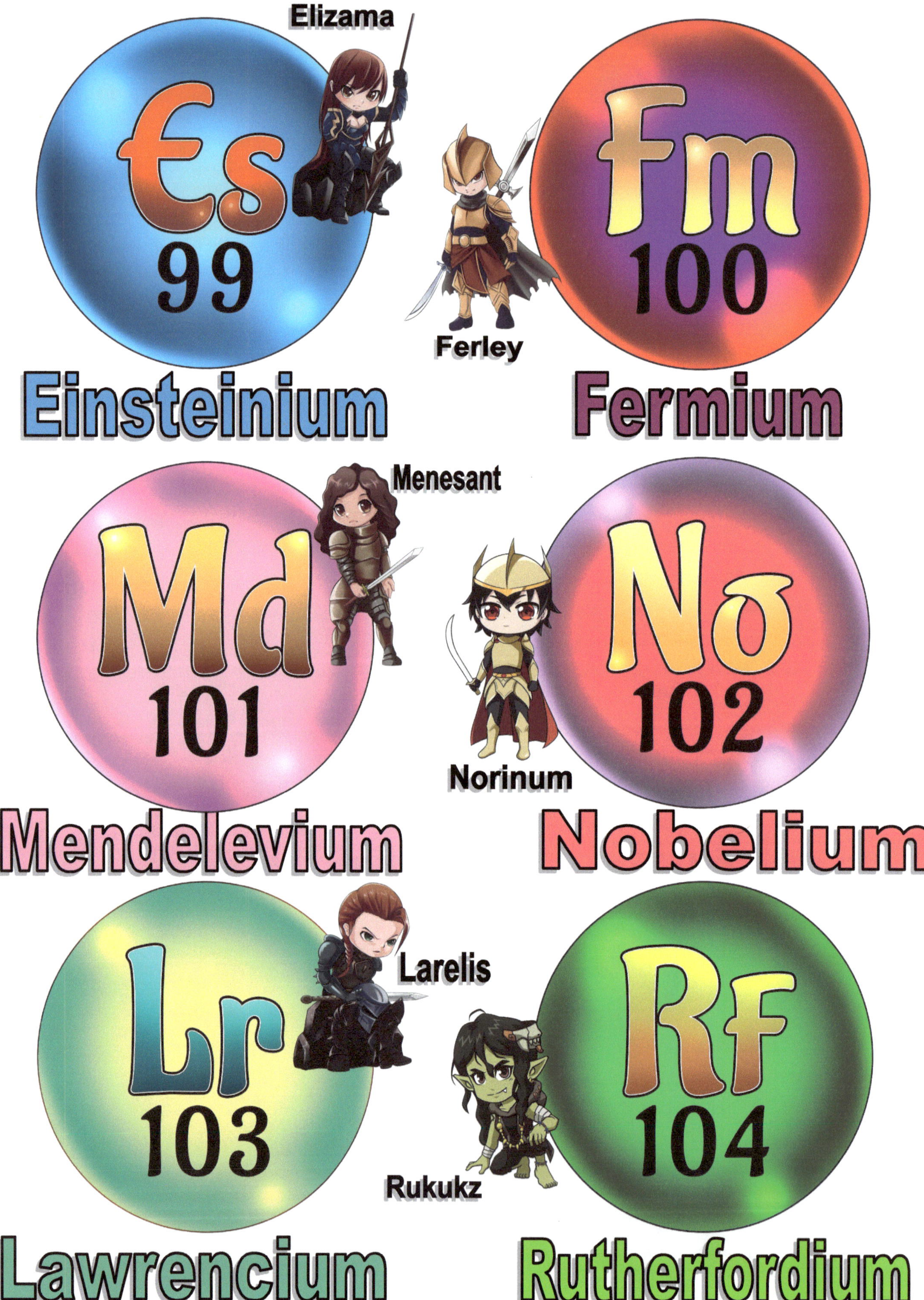

Elizama
Es
99
Einsteinium
Ferley
Fm
100
Fermium
Menesant
Md
101
Mendelevium
Norinum
No
102
Nobelium
Larelis
Lr
103
Lawrencium
Rukukz
Rf
104
Rutherfordium

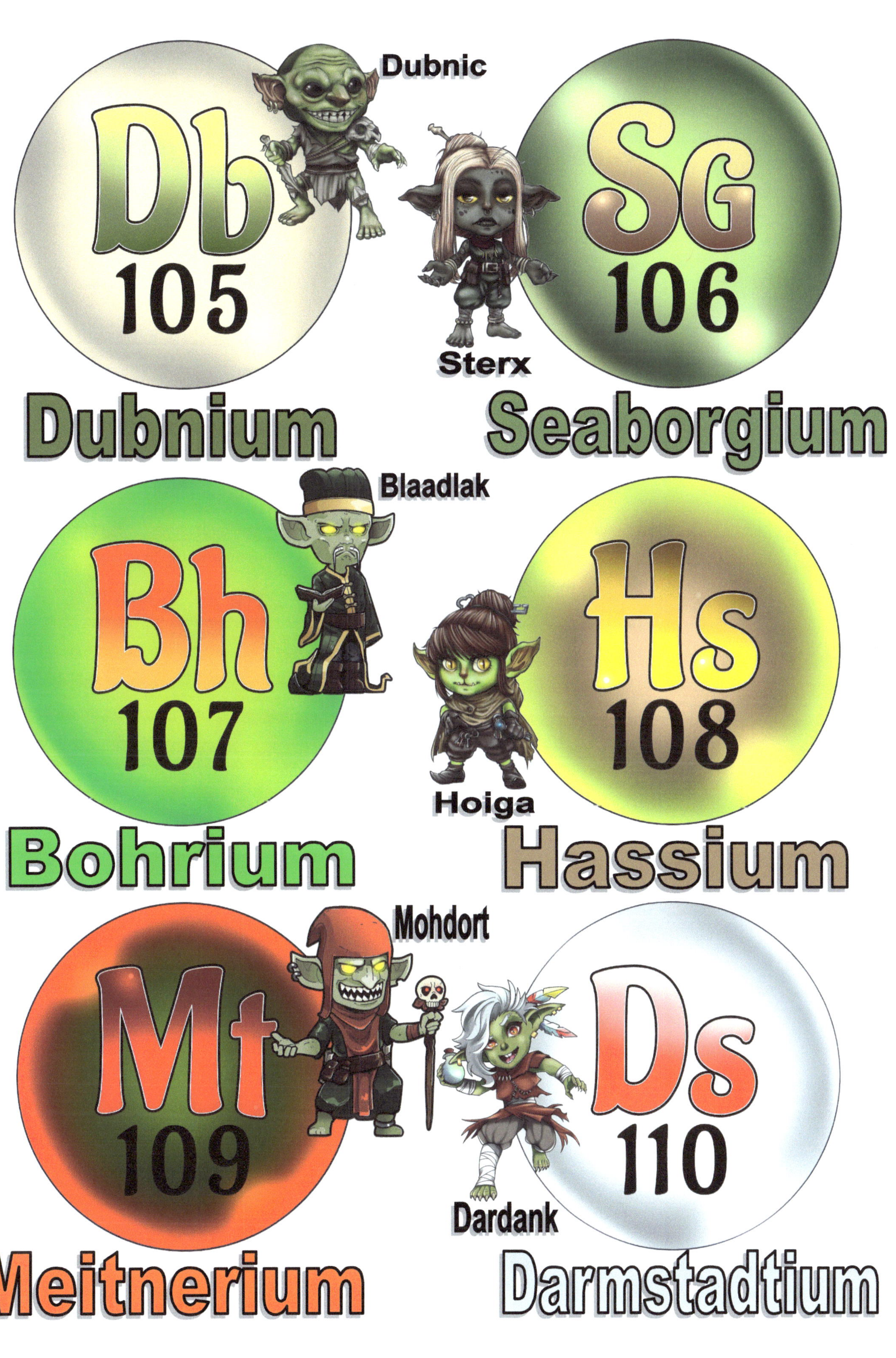

Dubnic
Db
105
Dubnium
Sterx
Sg
106
Seaborgium
Blaadlak
Bh
107
Bohrium
Hoiga
Hs
108
Hassium
Mohdort
Mt
109
Meitnerium
Dardank
Ds
110
Darmstadtium

Roentgenium

Copernicium

Nihonium

Flerovium

Moscovium

Livermorium

What Does The Future Hold?

So—What Comes Next For The Periodic Table?

There are currently 118 elements in the periodic table.

Of those, 28 elements are not found in nature. That means they were man-made by nuclear reactions and they have super short life spans, rendering them virtually unusable for anything other than scientific research.

Most of the elements with lower atomic numbers are found in nature but a very few of the elements with lower atomic numbers, such as Technetium (43), Promethium (61), Astatine (85) and Francium (87), are not naturally found in nature and are only produced in nuclear reactors.

Uranium (92) is generally considered the last natural element. Neptunium (93) and Plutonium (94) are generally synthetic, though trace amounts can occur naturally on earth.

All elements from Americium (95) onwards are synthetic and are known as the Transuranium Elements. Elements 104 and beyond have never been naturally found on earth.

Scientists keep trying to create new elements and in fact expect to increase the number on the periodic table to 137 or beyond. That's great for chemists working in laboratories but the big question for the rest of us is; will another natural element ever be discovered? It would seem that the search for all earthly elements has been exhausted but maybe not. The world is a very big place.

If such a discovery was made, where would it land on the periodic table? Would it be listed after the last used number or would it be squeezed in between it closest relatives? Might there be an element 25A someday in the deepest trenches of the ocean or buried under hundreds of layers of ice?

Or maybe, once humans start mining asteroids, completely new elements will surface that can be immediately used to create amazing things. What would their atomic numbers be? Would they deserve their own new series of numbers such as OE-1, indicating it is an Off Earth discovery?

The possibilities are endless that the best of humanity's discoveries are still waiting for us to find them.

There's A Dragon In My Atom!

Wondering why dragons, unicorns, wizards, knights, and goblins are hanging out with the atoms in this book? Welcome to the Magical Elementals version of the periodic table! Here, every element has a personality, a role to play, and a story to tell. You've just met all of the elementals who introduce the 118 amazing known elements in the *Magical Elements of the Periodic Table* six-book series.

Think of the periodic table as a vast fantasy world, where nature's elements and man-made elements are represented by magical elementals, each with powers inspired by the real-world wonders of the element they represent.

First we have Dragons, which have long captured our imaginations, often as creatures of immense power and mystery. The transition elements they represent are powerful, fierce, and impossible to ignore. Carbon is like a shape-shifting dragon—it can be soft as graphite or hard as diamond, and it's in almost everything. Hydrogen may be small, but it fuels the stars—a tiny dragon with cosmic fire. Lithium helps power our devices and calm our minds. Oxygen keeps us alive and fuels fire. Titanium is tough and light, perfect for planes and medical implants. These elements don't just exist—they change the world around them.

Next, we have the Wizards, who are often compared to the elements because, like fire, water, air, and earth, they are powerful, mysterious, and hard to contain. They tend to work alone and are often pushed to the edges of society. Some Wizards can be compared to noble gases like helium, neon, argon, krypton, xenon, and radon which stand quietly at the far right of the table. Just like wizards who keep to themselves, noble gases don't usually mix with others. They're calm and steady. But when they do get involved—like in neon signs or medical scans—they light up the world with brilliant glows. Only in rare moments do they join with other elements, much like wizards uniting for a secret, powerful ritual.

Now, say hello to the Unicorns. In a whimsical world, the elements of the periodic table can be imagined as unicorns, each with a distinct sparkle, power, and purpose. They can be compared to the transition metal elements which are beautiful, useful, and full of quiet magic. Gold shines bright and never tarnishes, like a timeless unicorn. Iron is strong and dependable—it's in our blood, our tools, and even the Earth's core. Copper conducts electricity and adds beauty to statues and buildings. Even tin, which may seem plain, helps preserve food and strengthen metals. These elements are essential, graceful, and full of purpose—each one adding something special to our world.

Together, the transition metals form a magical fellowship—diverse in color, strength, and spirit, yet united in their indispensable contributions to the system's balance and beauty. Next, meet the Knights, the brave warriors of the table. They're strong, a little unstable, and full of energy. Just like knights train and fight to become heroes, actinides like uranium and plutonium can release huge amounts of power. Both knights and elements show that strength often comes from interaction. They're used in things like cancer treatments and energy production. Their strength comes from change and teamwork— they're always transforming, just like a knight growing through battle and loyalty.

Finally, we have the Goblins, the superheavy radioactive elements. If wizards are noble gases and knights are actinides, then goblins—chaotic, unstable, shape-shifting, and lurking at the edges of understanding—are unmistakably the superheavy radioactive elements: bohrium, livermorium, oganesson, and their unstable kin at the far end of the periodic table. These are the tricksters. They don't stick around for long—they appear for just a moment before changing into something else. Unpredictable and hard to catch, they remind us that not everything in science (or in stories) is neat and tidy.

The best part? None of these "characters" work alone. Just like in any good fantasy tale, they need each other. Wizards may prefer solitude but sometimes they join forces. Knights fight together for a common purpose. Unicorns form fellowships that balance one another's gifts. Dragons, for all their power, are woven into the ecology of the worlds they inhabit. Even goblins, chaotic as they are, play a role — reminding every other creature that stability is earned, not guaranteed.

The periodic table isn't just a chart full of confusing symbols. It's a world of magic, power, and connection. Each element has its own personality, and when they come together, they create the world we live in—full of wonder, science, and story.

The Periodic Table Song

Beginning Chorus:

The Periodic Table
The Periodic Table
Electrons and Protons
Make each element stable
In the incredibly predictable
Elements
Of the Periodic Table.

Verse 1:

The Hydrogen atom has one of each.
That's what gives it the number one seat.

Helium is number two,
Two protons give the sky its blue.

Lithium with three electrons to share,
Third in line, it's light as air.

Beryllium comes right after that,
Four protons strong—imagine that.

Boron's five, it's next in line,
Five electrons, oh so fine.

Repeat Chorus:

The Periodic Table
The Periodic Table
Electrons and Protons
Make each element stable
In the incredibly predictable
Elements
Of the Periodic Table.

Bridge

Every element has its place
On the big periodic space.
Each one has a number, too,
Tells you what its atoms do.

Count them all and you will see,
The wonders of chemistry.

Verse 2:

Carbon has six, it's life's best friend,
It helps all living things to mend.

Nitrogen atom seven protons there.
Seven electrons make the air we share.

Oxygen's eight for the fires glow,
And it makes our bodies run just so.

Fluorine is nine, a real socialite.
It loves to bond with all its might.

The Neon atom ten protons bright.
Ten electrons make signs glow at night.

Repeat Chorus:

The Periodic Table
The Periodic Table
Electrons and Protons
Make each element stable
In the incredibly predictable

The Periodic Table
The Periodic Table
Electrons and Protons
Make each element stable
In the incredibly predictable
Elements
Of the Periodic Table.

Watch the video at

https://www.youtube.com/watch?v=x6npdgY57C8

Dear Reader

I hope the Magical Elemental Atoms—Count The Protons and Electrons book has helped you learn a little something about the magic of the periodic table.

The artwork was created by Pranavva (@pranavva_ on Instagram & @nearwasa on Fiverr).

This book compliments the Magical Elements of the Periodic Table book series which features elementals introducing chemical elements. Every one of them is amazing and very necessary to our everyday lives. All of the elements are truly

Techno-magical.

A lot of research went into every page of this book as well as all the other Magical Elements of the Periodic Table Books. There are just too many references to publish in this book but you can read and research them all at MagicalPTElements.com/MAUPT or /MDAPT or /MW1PT or /MW2PT or /MAKPT or /MRGPT. There, you can also access book related activity sheets, coloring pages, games and even Elemental tee-shirts to help make the learning process more fun. You can see a few fun video shorts there, too.

Get ready-made trading cards, lapel pins, tee shirts and more based on this book from Sybrina Publishing's No Metal No Magic Collection at Zazzle - **http://bit.ly/3km64Wg**

Would you like a 24" x 36" poster of the Elemental-Themed Periodic Table in this book? The best place to get it is at **https://bit.ly/49QMxBT** They have the sharpest images of any other poster printer around.

The Magical Elements of the Periodic Table books came into existence because of

my Blue Unicorn—Journey To Osm books. If it weren't for their magical powers,

based on the properties of the metals of their horns and hooves, I would have

never come up with the idea to relate magical creatures to the periodic table.

There's a metal horn unicorn story for every age group and they are all available

at Sybrina's Unicorn Book Store on at Sybrina.com.

If you enjoyed this book
please leave a nice review
at your favorite online book site.

Do Your Middle Graders Want To Know More From The Magical Elementals About The Periodic Table?

Get the accompanying books in print at all online book stores. Get the books and accompanying activities at MagicalPTElements.

Available Now

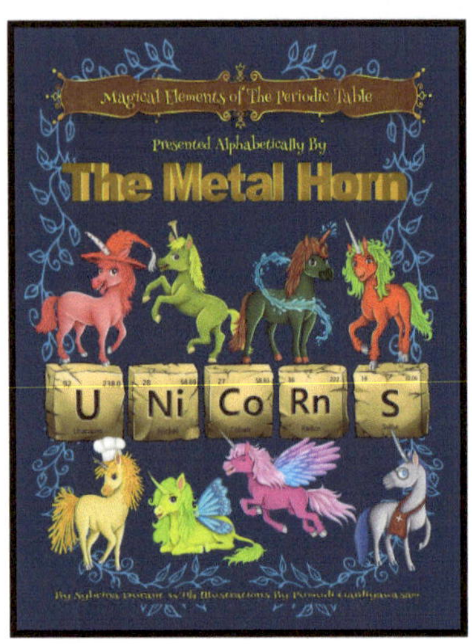

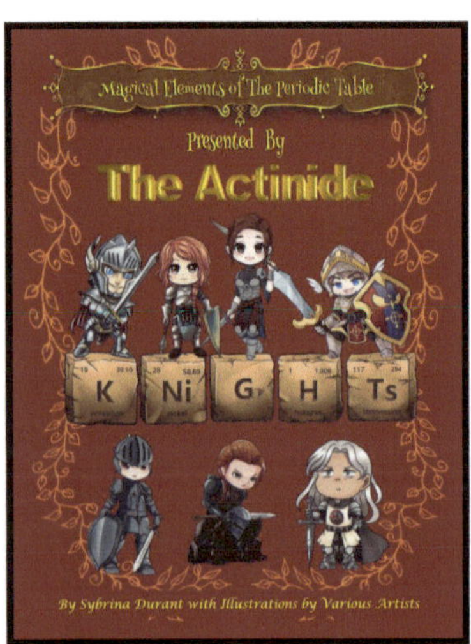